Localized Carbon Sequestration Success Stories

Cassia Parham

Copyright © 2024 Crafted Spaces LLC

All rights reserved.

Dedication

To my incredible family, friends, and mentors—your unwavering support and belief in my vision has been the driving force behind this book. Your encouragement, wisdom, and love have made this journey possible, and I am endlessly grateful. I dedicate this work to you and to every individual who is striving to make the world a better place, one small, meaningful action at a time. May this book inspire you as much as your dedication to sustainability has inspired me. Together, we can create lasting change.

Chapter 1 6
    Introduction to Carbon Sequestration 6
Chapter 2 14
    Understanding Localized Approaches 14
Chapter 3 21
    Agricultural Practices for Carbon Sequestration 21
Chapter 4 29
Urban Carbon Sequestration Initiatives 29
Chapter 5 37
    Restoration of Natural Ecosystems 37
Chapter 6 44
    Innovative Technologies for Localized Sequestration
    44
Chapter 7 52
    Policy Frameworks Supporting Local Initiatives 52
Chapter 8 59
    Community Engagement and Education 59
Chapter 9 65
    Measuring and Verifying Carbon Sequestration 65
Chapter 10 72
    Challenges and Future Directions 72
Conclusion 80

- The Path Forward 80
- Frequently Asked Questions 89
- Case Study Summary 97
- Resources 104
- Glossary 112
- References 119
- About the Author 126

Chapter 1
Introduction to Carbon Sequestration

Climate change is one of the most pressing challenges of our time, and **carbon sequestration** is a critical tool in our efforts to mitigate its impact. Carbon sequestration involves capturing and storing atmospheric carbon dioxide (CO_2) to reduce the amount of this potent greenhouse gas in the atmosphere. By storing CO_2 in soil, plants, geological formations, or even the ocean, carbon sequestration helps slow the accumulation of CO_2 that drives global warming. This chapter explores the significance of carbon sequestration, its underlying mechanisms, and the potential impact of its widespread implementation.

Defining Carbon Sequestration: Understanding Its Role in Climate Change Mitigation

Carbon sequestration plays a pivotal role in mitigating climate change. As human activities—like burning fossil fuels—release vast amounts of CO2 into the atmosphere, the greenhouse effect intensifies, trapping heat and leading to global temperature rise. Carbon sequestration offsets these emissions by removing CO2 from the atmosphere and storing it in various forms.

This process can take place through natural systems, such as **biological sequestration** in forests, soils, and oceans, or through engineered solutions like **geological sequestration**, which captures CO2 from industrial sources and stores it deep underground. Both methods aim to lower the concentration of CO2 in the atmosphere, reducing the intensity of the greenhouse effect (Lal, 2004; Paustian et al., 2019).

Basic Principles of Carbon Capture and Storage (CCS): How CCS Works and Why It Matters

Carbon capture and storage (CCS) is a technological process designed to capture CO_2 emissions from industrial activities and power plants before they enter the atmosphere. The process compresses and transports the captured CO_2—often via pipelines—to underground reservoirs, where it safely stores the carbon for long periods (Bachu, 2008). This process typically involves three steps:

1. **Capturing CO_2** from large sources like factories and power plants.

2. **Transporting** the captured CO_2 through pipelines or ships.

3. **Storing** CO2 in geological formations such as depleted oil fields or saline aquifers (Benson & Cole, 2008).

CCS is essential for industries that are difficult to decarbonize, such as cement and steel production, and can significantly reduce emissions from these sectors. By implementing CCS at scale, we could achieve substantial reductions in global CO2 emissions, making it a crucial technology in our fight against climate change (Lal, 2004).

Geological vs. Biological Sequestration: Key Differences and Applications

There are two primary methods of carbon sequestration: **geological** and **biological**.

Geological sequestration stores CO_2 in underground rock formations, preventing it from contributing to the greenhouse effect. This method is typically used to store CO_2 from industrial sources and has a high potential for storing vast amounts of carbon (Benson & Cole, 2008). Geological storage sites, such as depleted oil and gas reservoirs or deep saline aquifers, can securely hold CO_2 for thousands of years (Bachu, 2008).

Biological sequestration relies on natural processes to absorb CO_2. Plants, trees, and soils capture CO_2 through photosynthesis, storing it in biomass and the earth. Forests, wetlands, and agricultural soils are key players in biological sequestration, as they act as natural carbon sinks (Paustian et al., 2019). While biological sequestration applies more widely and benefits ecosystems, it works more slowly and stores carbon less permanently than geological methods, as plants or soils can release carbon back into the atmosphere through deforestation, land-use changes, or natural disturbances. (Lal, 2004).

Global Impact: How Carbon Sequestration Could Reduce Emissions by 15%

The global potential for carbon sequestration is immense. Research suggests that if effectively implemented across various sectors, carbon sequestration could reduce global greenhouse gas emissions by as much as 15% by 2050 (International Energy Agency, 2020). This reduction is vital to achieving international climate goals, such as limiting global warming to 1.5°C above pre-industrial levels, as outlined in the Paris Agreement (IEA, 2020).

While both geological and biological sequestration methods offer promising solutions, a combination of the two will be necessary to meet the scale of the challenge. To realize this potential, countries must invest in CCS infrastructure, reforestation, and sustainable agricultural practices, while also enacting supportive policies to encourage widespread adoption (Paustian et al., 2019).

Conclusion

Carbon sequestration is a critical strategy in the global effort to combat climate change. By capturing and storing CO_2 through both technological and natural methods, we can significantly reduce the amount of carbon in the atmosphere, helping to slow the rate of global warming. Geological sequestration provides a secure, long-term solution for industrial emissions, while biological sequestration enhances natural carbon sinks. Together, these methods have the potential to reduce emissions by

15%, making carbon sequestration an essential tool in our path toward a sustainable future.

Key Takeaway

Carbon sequestration, whether through natural or engineered means, is a powerful approach to reducing atmospheric CO_2 and mitigating climate change. When widely implemented, it can play a pivotal role in achieving global climate goals.

Chapter 2

Understanding Localized Approaches

Addressing climate change requires innovative, localized solutions tailored to specific ecosystems and communities. Carbon sequestration offers a valuable tool for this, but the success of these efforts relies on community involvement and region-specific strategies. In this chapter, we explore the importance of localized carbon sequestration, discuss how tailored methods can boost effectiveness, and examine real-world success stories from the Midwest U.S. and rural India, where community-driven approaches have led to significant improvements in soil carbon levels.

The Importance of Local Initiatives:

Why Community-Level Efforts Are Crucial

Localized carbon sequestration initiatives are essential because they directly engage communities in climate action while responding to specific environmental conditions. Local initiatives allow for a deep understanding of regional ecosystems and agricultural practices, fostering a sense of ownership and responsibility within the community. By empowering local actors, these projects can lead to more sustainable, long-lasting results (Paustian et al., 2019).

Communities in different regions face unique challenges. Whether it's soil erosion in the Midwest or land degradation in rural India, localized approaches allow for solutions that fit specific conditions, making carbon sequestration more effective. These efforts can contribute to broader environmental and social benefits, such as improved soil health, biodiversity, and economic resilience (Lal, 2004).

Tailored Methods for Specific Environments: Customizing Strategies for Different Regions

One of the primary benefits of localized approaches to carbon sequestration is the ability to tailor methods to suit the environment. In regions like the Midwest U.S., agricultural practices such as **no-till farming** and **cover cropping** have proven highly effective in improving soil health and sequestering carbon. No-till farming leaves the soil undisturbed, preserving its carbon stores and preventing erosion, while cover crops improve soil structure and organic matter content, boosting its carbon sequestration potential (Reeves, 1997).

In contrast, rural India has seen success through **agroforestry**, which integrates trees with agricultural land to sequester carbon while providing shade and resources for crops. This method works particularly well in tropical and semi-arid regions, where deforestation and soil degradation are major challenges. By planting trees that absorb CO_2 and support crop growth, farmers can simultaneously capture carbon and improve the productivity of their land (Rao et al., 2007). Tailoring these approaches to fit local conditions ensures that carbon sequestration efforts are both practical and impactful.

Case Studies: Midwest U.S. and Rural India Successes in Enhancing Soil Carbon by 20%

In the **Midwest U.S.**, a combination of no-till farming and cover cropping has led to a remarkable increase in soil organic carbon. Over a five-year period, farmers practicing these techniques have reported a **20% increase in soil carbon levels**, demonstrating the power of localized agricultural practices in sequestering carbon and improving soil health (Lal, 2004). By reducing soil disturbance and maintaining continuous ground cover, Midwest farmers have prevented carbon loss while enhancing water retention and crop yields.

Similarly, in **rural India**, the implementation of agroforestry has yielded impressive results. In some regions, soil carbon levels have increased by **20%** over the same five-year period, thanks to the integration of trees and crops (Rao et al., 2007). This approach has not only sequestered carbon but also helped restore degraded land, increased biodiversity, and improved the livelihoods of local farmers. These case studies highlight the importance of adapting carbon sequestration techniques to fit the specific environmental and socio-economic contexts of each region.

Conclusion

Localized approaches to carbon sequestration offer a powerful solution to climate change by aligning with the unique environmental and community needs of each region. Whether it's through no-till farming and cover cropping in the Midwest U.S. or agroforestry in rural India, these methods have proven effective in boosting soil carbon levels by as much as 20%. By empowering communities and tailoring strategies to local conditions, carbon sequestration efforts can contribute to both climate mitigation and improved agricultural productivity.

Key Takeaway

Localized carbon sequestration initiatives, tailored to specific environments and communities, are crucial for maximizing the effectiveness of carbon capture and improving soil health. Success stories from the Midwest U.S. and rural India demonstrate the power of community-driven, region-specific strategies in achieving significant results.

Chapter 3
Agricultural Practices for Carbon Sequestration

Agriculture plays a crucial role in mitigating climate change by capturing and storing carbon dioxide (CO_2) in the soil. By implementing innovative farming techniques, farmers can actively contribute to carbon sequestration while improving soil health and crop yields. This chapter explores three key practices: no-till farming, cover cropping, and agroforestry. Through a case study of Brazilian farmers, we examine how these methods have led to a significant increase in soil organic carbon (SOC) levels, demonstrating the potential for sustainable agriculture to address both environmental and economic challenges.

Innovative Farming Techniques:

Exploring No-Till Farming, Cover Cropping, and Agroforestry

No-till farming is one of the most effective conservation practices for carbon sequestration. Unlike traditional tilling methods, which disturb the soil and release stored carbon, no-till farming leaves the soil undisturbed, allowing carbon to remain sequestered. By reducing soil disturbance, farmers preserve the integrity of the soil structure and foster the growth of microorganisms that play a key role in carbon storage (Herbert et al., 2005). No-till farming also enhances water retention and reduces soil erosion, making it an ideal practice for regions prone to drought or heavy rainfall.

Cover cropping involves planting crops such as clover, rye, or legumes during the off-season when the soil would otherwise be left bare. These cover crops protect the soil from erosion, add organic matter, and improve nutrient cycling. By maintaining year-round ground cover, farmers can prevent the loss of carbon from the soil and increase its organic matter content. Cover crops also support the microbial communities in the soil, which further aids in carbon sequestration (Reeves, 1997).

Agroforestry integrates trees and shrubs into traditional farming systems, providing a dual benefit of carbon capture and agricultural productivity. Trees naturally sequester carbon through photosynthesis, storing it in their biomass and roots. When combined with crops, agroforestry systems create a diverse and resilient ecosystem that enhances soil health, reduces the need for chemical fertilizers, and provides additional sources of income for farmers through timber or fruit production (Ramos et al., 2018). Agroforestry is especially beneficial in tropical regions where deforestation and land degradation are major challenges.

Case Study: Brazilian Farmers Increasing Soil Organic Carbon by 1.5% Annually

Brazil provides a compelling example of how these agricultural practices enhance carbon sequestration. In southern Brazil, farmers have adopted no-till farming and agroforestry techniques, leading to significant increases in soil organic carbon. Over five years, these practices have resulted in an annual SOC increase of **1.5%**. The combination of leaving the soil undisturbed and integrating trees into crop systems has not only sequestered carbon but also improved soil fertility and reduced erosion (Ramos et al., 2018). Brazilian farmers have become global leaders in sustainable agriculture by demonstrating that carbon sequestration and agricultural productivity can go hand in hand.

The success of these techniques is notable in Brazil's Paraná region, where no-till farming has reduced soil erosion by up to 80%. These farmers protect the soil structure and enhance its ability to store carbon, creating a sustainable model that others can replicate in regions facing similar environmental challenges (Manlay et al., 2007).

Dual Benefits: Carbon Capture and Improvements in Soil Health and Crop Yields

The dual benefits of these agricultural practices extend beyond carbon capture. By increasing SOC levels, farmers are enhancing the overall health of their soil, leading to higher crop yields and greater resilience to climate variability. Healthier soil retains more water, reduces the need for irrigation, and provides a stable environment for crops to thrive. Additionally, reducing the use of chemical fertilizers through methods like agroforestry and cover cropping lowers input costs and decreases the environmental impact of farming (Oldfield et al., 2019).

Increased soil carbon also supports biodiversity, as healthy soils are home to a wide range of microorganisms that contribute to nutrient cycling and pest control. This natural pest management reduces the need for pesticides, further enhancing the sustainability of farming systems. Overall, these practices offer a sustainable way to improve agricultural productivity while addressing the urgent need to reduce carbon emissions.

Conclusion

Agricultural practices such as no-till farming, cover cropping, and agroforestry offer powerful tools for carbon sequestration. The case of Brazilian farmers increasing soil organic carbon by 1.5% annually serves as a testament to the potential of these techniques to mitigate climate change while improving soil health and crop yields. As farmers around the world adopt these practices, they not only contribute to global carbon reduction efforts but also create more resilient and sustainable agricultural systems.

Key Takeaway

Innovative farming techniques like no-till farming, cover cropping, and agroforestry not only sequester carbon but also enhance soil health, increase crop yields, and reduce environmental impact, making them essential tools for addressing climate change through agriculture.

Chapter 4

Urban Carbon Sequestration Initiatives

As cities grow and become more densely populated, they face significant environmental challenges. Urban areas, traditionally seen as carbon sources because of heavy traffic and industrial activity, now have the potential to play a crucial role in mitigating climate change through innovative carbon sequestration strategies. In this chapter, we explore urban sequestration techniques, such as green roofs, urban forests, and community gardens. We also examine Melbourne's successful urban greening projects, which have sequestered hundreds of thousands of tons of

CO2, while providing additional benefits such as improved biodiversity and air quality.

Urban Sequestration Strategies:

Green Roofs, Urban Forests, and Community Gardens

Urban environments provide unique opportunities for carbon sequestration by incorporating nature-based solutions into the city infrastructure. **Green roofs**, for example, turn otherwise barren rooftop spaces into thriving ecosystems that absorb CO2, reduce heat, and improve insulation, thus lowering energy consumption. By covering rooftops with plants, cities can create a cooler, more pleasant environment while capturing carbon and offsetting emissions (Taghi Miri et al., 2024).

Urban forests are another impactful method of carbon capture. These tree-filled spaces provide a sanctuary for wildlife while sequestering carbon. Trees absorb CO2 through photosynthesis, and when strategically planted in cities, they also help reduce the urban heat island effect by offering shade and cooling the air. In Melbourne, for instance, the city's Urban Forest Fund has supported the planting of thousands of trees, enhancing green spaces and increasing the city's capacity to capture and store carbon (City of Melbourne, 2021).

Community gardens play a dual role in urban carbon sequestration by promoting sustainable food production and improving soil carbon storage. These gardens, often run by local volunteers, provide green space in densely populated areas, supporting plant growth and soil health, which captures carbon. By fostering a sense of community, they also encourage environmental stewardship at the local level (City of Melbourne, 2021).

Case Study: Melbourne's Urban Greening Projects Sequestering 300,000 Tons of CO2

Melbourne has emerged as a leader in urban carbon sequestration, with its **Urban Forest Strategy** aiming to transform the city into a green urban oasis. By investing in large-scale planting initiatives and green infrastructure, Melbourne has sequestered approximately **300,000 tons of CO2** over the past decade (Living Melbourne, 2021). These efforts include the development of urban forests, green roofs, and the rehabilitation of degraded urban spaces.

One standout project is the **Greening the West** initiative, which has planted over a million trees across Melbourne's western suburbs. This initiative not only increases the region's carbon capture capacity but also enhances urban biodiversity, reduces the urban heat island effect, and creates healthier, more livable spaces for residents (Greening the West, 2021). Additionally, Melbourne's **Skyfarm** project—a rooftop farm—has contributed to urban carbon sequestration by transforming concrete rooftops into green spaces that absorb CO_2 while producing food for local communities (City of Melbourne, 2021).

Additional Benefits: Biodiversity and Air Quality Improvements Alongside Carbon Capture

Urban carbon sequestration efforts bring a host of additional benefits beyond carbon capture. One of the most significant is the improvement of **biodiversity**. Green spaces in urban areas create habitats for a variety of plant and animal species, fostering ecosystems that are often absent in concrete-dominated cities. For example, small greening projects in Melbourne have led to a sevenfold increase in insect species, which play a vital role in pollination and ecosystem health (City of Melbourne, 2021).

Moreover, urban greening improves **air quality** by filtering pollutants such as particulate matter and nitrogen dioxide. Plants absorb these harmful substances, making the air cleaner and healthier for urban populations. This, combined with the cooling effect of green spaces, helps to mitigate some of the negative impacts of climate change, such as heatwaves and poor air quality (Taghi Miri et al., 2024).

Conclusion

Urban carbon sequestration is a vital component of global climate action. By incorporating green roofs, urban forests, and community gardens into city planning, urban areas can not only reduce their carbon footprint but also improve biodiversity, air quality, and overall livability. Melbourne's success in sequestering 300,000 tons of CO_2 through its urban greening projects provides a model for other cities to follow. As more urban areas adopt similar strategies, they can become powerful forces for carbon capture and environmental sustainability.

Key Takeaway

Urban carbon sequestration initiatives, such as green roofs, urban forests, and community gardens, offer cities the opportunity to capture carbon while improving biodiversity, air quality, and overall livability, as demonstrated by Melbourne's greening projects.

Chapter 5
Restoration of Natural Ecosystems

Natural ecosystems such as wetlands, forests, and grasslands play a crucial role in sequestering carbon and mitigating climate change. These ecosystems act as carbon sinks, absorbing and storing CO_2 from the atmosphere through vegetation and soil processes. This chapter explores the restoration of these ecosystems, with a focus on rewilding initiatives in the Scottish Highlands, which have sequestered millions of tons of CO_2, and the integration of Indigenous knowledge in maintaining ecosystem resilience and carbon storage.

Restoring Wetlands, Forests, and Grasslands: The Role of Natural Ecosystems in Carbon Capture

Restoring natural ecosystems is vital for increasing global carbon sequestration. **Wetlands** are some of the most efficient ecosystems for carbon storage. These waterlogged environments prevent the decomposition of organic matter, locking carbon in the soil for centuries. As wetlands are restored, they can absorb and store large quantities of CO2, acting as natural buffers against climate change.

Forests, especially those rich in biodiversity, are another key carbon sink. Through photosynthesis, trees absorb CO2 from the air and store it in their biomass and soil. When restored, forests not only capture more carbon but also improve biodiversity and ecosystem resilience. **Grasslands** also sequester significant amounts of carbon in their root systems. When degraded grasslands are restored, their deep-rooted plants help to trap carbon in the soil, preventing its release into the atmosphere (Lal, 2004).

Case Study: Rewilding the Scottish Highlands with 5 Million Tons of CO2 Sequestered

One of the most remarkable examples of ecosystem restoration is the rewilding of the **Scottish Highlands**. This large-scale project aims to restore the Caledonian Forest, which once covered much of the region, to its former glory. By planting over 1.5 million native trees and promoting natural regeneration through controlled deer populations, the rewilding project has already sequestered an estimated **5 million tons of CO2** (Rewilding Europe, 2021).

The initiative, led by organizations such as Trees for Life, focuses on connecting fragmented habitats across the Highlands to create a vast, interconnected network of rewilded landscapes. This not only sequesters carbon but also boosts biodiversity, creating habitats for species like red squirrels and beavers, which conservationists have reintroduced as part of the effort. The project closely ties to local communities, providing green jobs and promoting eco-tourism (Highlands Rewilding, 2021).

Indigenous Knowledge: Integrating Traditional Practices for Ecosystem Resilience and Carbon Storage

Indigenous communities have long understood the value of maintaining healthy ecosystems for both environmental and cultural reasons. Their traditional practices often emphasize **sustainable land management** that enhances the natural carbon sequestration potential of ecosystems. For instance, Indigenous fire management techniques, such as controlled burns, help prevent large-scale wildfires and maintain the health of grasslands and forests, improving their ability to store carbon.

Incorporating **Indigenous knowledge** into modern conservation efforts can significantly improve the effectiveness of ecosystem restoration. These practices prioritize ecological balance and biodiversity, ensuring that ecosystems remain resilient in the face of climate change. By partnering with Indigenous communities, rewilding projects can benefit from centuries of land stewardship experience, creating more sustainable and effective carbon sequestration strategies (IPCC, 2019).

Conclusion

Restoring natural ecosystems is one of the most effective ways to combat climate change through carbon sequestration. Wetlands, forests, and grasslands play a critical role in capturing and storing carbon, and their restoration can have far-reaching environmental benefits. The rewilding of the Scottish Highlands shows the immense potential for ecosystem restoration, with millions of tons of CO2 already sequestered. Integrating Indigenous knowledge into these efforts ensures the long-term sustainability and resilience of restored ecosystems.

Key Takeaway

Restoring wetlands, forests, and grasslands through initiatives like the rewilding of the Scottish Highlands is essential for enhancing carbon sequestration and preserving biodiversity. Integrating Indigenous knowledge into these efforts adds invaluable ecological insights for building resilient ecosystems.

Chapter 6

Innovative Technologies for Localized Sequestration

As the need to combat climate change becomes more urgent, innovative technologies are emerging to help capture and store atmospheric carbon. Among these, **Bioenergy with Carbon Capture and Storage (BECCS)** and **Direct Air Capture (DAC)** are at the forefront of global efforts to remove CO2 from the atmosphere. This chapter explores these advanced technologies, highlights companies investing in their development, and examines the future scalability of these solutions.

Exploring Advanced Sequestration Technologies: BECCS and Direct Air Capture (DAC)

Bioenergy with Carbon Capture and Storage (BECCS) combines the use of bio-energy (from organic materials) with carbon capture technologies to remove CO_2 from the atmosphere. In this process, biomass generates energy, and the system captures and stores the CO_2 emitted during combustion underground, effectively reducing atmospheric carbon levels. BECCS offers a dual benefit: generating renewable energy while simultaneously sequestering carbon. According to the International Energy Agency (IEA), BECCS could sequester between 0.5 and 5 billion tons of CO_2 annually if implemented on a large scale (International Energy Agency, 2020).

Direct Air Capture (DAC), involves extracting CO_2 directly from the air using chemical processes. This CO_2 is then either stored underground or used in various industries, such as for producing synthetic fuels. DAC has the potential to be deployed anywhere, making it a versatile solution for addressing emissions in locations that lack other natural carbon sinks, such as forests or grasslands. Companies like **Climeworks** and **Heirloom** are pioneering DAC technologies and are actively working to scale their operations (Carbon Credits, 2024). Climeworks' **Mammoth** facility in Iceland, the largest of its kind, currently captures 4,000 tons of CO2 annually, and its upcoming projects in the U.S. aim to scale this capacity even further (S&P Global, 2024).

Case Studies: Companies Investing in Cutting-Edge Tech to Remove 1 Billion Tons of CO2 Annually

Several companies are making significant investments in BECCS and DAC technologies to help meet global carbon removal targets. **Occidental Petroleum's subsidiary, 1PointFive**, is building one of the largest DAC facilities in Texas. **The South Texas DAC Hub**, part of a $1.2 billion federal initiative, expects to remove up to 1 million tons of CO2 annually once it becomes operational by 2025 (Carbon Herald, 2023). This project exemplifies the growing commitment from both private and public sectors to scale up DAC technology.

Meanwhile, **Drax Group**, a leader in BECCS, is working to develop large-scale bio-energy carbon capture projects. Drax's ongoing BECCS initiatives expect to remove millions of tons of CO2 by capturing emissions from its biomass energy plants. By converting biomass into energy and capturing the resulting carbon, Drax is simultaneously helping reduce emissions and generating renewable energy (IEA, 2020).

Together, these companies aim to remove over 1 billion tons of CO2 annually as their projects come online, marking a significant step toward achieving global climate goals.

Technological Potential:

Future Projections and Scalability

The potential of BECCS and DAC technologies is vast. By 2050, experts estimate that **up to 10 billion tons of CO2** must be removed from the atmosphere each year to keep global temperature rise below 1.5°C, a target set by the Paris Agreement (BCG, 2024). Achieving this scale will require significant advancements in both technologies and cost reductions. Currently, the cost of DAC is relatively high—around $200 to $600 per ton of CO_2—but experts project costs could fall to **$150 per ton or less** as the technology matures (S&P Global, 2024).

Both BECCS and DAC are poised for dramatic expansion. BECCS, with its dual benefit of energy production and carbon removal, offers an immediate pathway to scale up through existing biomass infrastructure. DAC, while more energy-intensive, has the advantage of flexibility in terms of location, allowing it to be deployed in areas lacking natural carbon sinks (BCG, 2024). With increased investment and technological innovation, both BECCS and DAC could become essential components of the global carbon removal strategy, helping to offset emissions from hard-to-decarbonize industries like aviation and heavy manufacturing.

Conclusion

Innovative technologies like BECCS and DAC represent the future of carbon sequestration. With companies making large-scale investments in these technologies and governments providing crucial funding, we are on the cusp of a carbon removal revolution. As costs decrease and scalability increases, these technologies have the potential to remove **up to 10 billion tons of CO2** annually by mid-century, making a significant contribution to global efforts to combat climate change.

Key Takeaway

Bioenergy with Carbon Capture and Storage (BECCS) and Direct Air Capture (DAC) are at the forefront of innovative carbon sequestration efforts. As these technologies scale, they offer the potential to remove billions of tons of CO2 from the atmosphere, playing a critical role in achieving global climate targets.

Chapter 7
Policy Frameworks Supporting Local Initiatives

Effective climate action often requires strong policy frameworks to support local carbon sequestration initiatives. Governments play a critical role in incentivizing sustainable practices and ensuring that communities and industries have the resources and guidance to reduce their carbon footprints. In this chapter, we explore how well-crafted policy models, like the European Union's Green Deal, are driving carbon sequestration at the local level, discuss a case study from the EU, and analyze the success of these policies in increasing sequestration activities by 30% over the past five years.

Effective Policy Models: How Governments Incentivize Carbon Sequestration at Local Levels

Governments around the world are increasingly implementing policies that encourage local carbon sequestration efforts. These policies provide financial incentives, regulatory support, and technical resources to help communities and industries adopt practices that capture and store carbon. For example, through **subsidies, tax credits, and grants**, governments can encourage the use of carbon capture technologies in agriculture, forestry, and industrial sectors. These initiatives ensure that local actors are not burdened by the financial costs of implementing sustainable practices.

A successful example is the EU's **Common Agricultural Policy (CAP)**, which has increasingly integrated sustainability measures that reward farmers for adopting carbon sequestration techniques like agroforestry and regenerative agriculture (European Commission, 2023). Through this policy, local farmers receive financial support to transition toward practices that store carbon in the soil, improving soil health while contributing to emissions reduction goals.

Case Study: The EU's Green Deal and Its Impact on Carbon Farming and Sustainable Investments

The **European Green Deal**, launched in 2019, is one of the most ambitious policy frameworks designed to make Europe the first carbon-neutral continent by 2050. This plan includes a comprehensive package of regulatory and financial measures aimed at reducing net greenhouse gas emissions by **at least 55% by 2030**, compared to 1990 levels (European Commission, 2021). One of the key components of this framework is **carbon farming**, a concept that encourages farmers to adopt practices that sequester carbon in agricultural land. These efforts are supported by subsidies and access to sustainable investment funds, which provide the necessary resources to make carbon sequestration economically viable at the local level.

Through the Green Deal's **Sustainable Europe Investment Plan**, the EU has mobilized **€1 trillion in green investments** to support projects ranging from carbon capture in agriculture to renewable energy production (European Commission, 2021). This level of financial commitment has made a tangible impact: over the past five years, the EU has seen a **30% increase** in carbon sequestration activities, particularly in agriculture and forestry, thanks to these strategic investments (World Economic Forum, 2023).

Policy Success: A 30% Increase in Sequestration Activities Over the Last Five Years

The significant increase in local carbon sequestration activities can measure the success of the EU's Green Deal policies. By creating financial incentives for carbon farming and other sustainable practices, and by investing in infrastructure that supports carbon capture, the EU has showed the effectiveness of well-designed policies in driving environmental change. According to recent reports, sequestration activities across the EU have increased by **30%** over the last five years, particularly in sectors like agriculture, forestry, and energy (World Economic Forum, 2023).

The **EU Emissions Trading System (ETS)** has played a critical role in this progress by putting a price on carbon emissions. Companies are incentivized to reduce their emissions through either direct reductions or by investing in carbon offset projects, many of which involve reforestation or other land-use practices that capture and store carbon. This market-based approach has contributed significantly to the overall increase in carbon sequestration efforts, further solidifying the success of the EU's policy framework (European Commission, 2021).

Conclusion

Effective policy frameworks like the EU's Green Deal show the power of government action in supporting local carbon sequestration initiatives. By providing financial resources, regulatory support, and incentives for sustainable practices, governments can empower communities and industries to adopt carbon capture technologies. The success of the Green Deal, with its 30% increase in sequestration activities over the last five years, highlights the importance of continued investment and regulatory support to achieve long-term climate goals.

Key Takeaway

Strong policy frameworks, such as the EU's Green Deal, are essential for promoting local carbon sequestration initiatives. Through financial support and regulatory incentives, these policies have led to a significant increase in sequestration activities, helping to drive the global effort to combat climate change.

Chapter 8

Community Engagement and Education

In the fight against climate change, engaging local communities is essential to the success of carbon sequestration projects. Building awareness and fostering a sense of ownership over these initiatives ensures communities implement and sustain them. This chapter explores the role of community engagement in promoting carbon sequestration, highlights Kenya's success in training thousands of farmers in sustainable practices, and discusses the power of education in driving community-led climate action.

The Role of Local Communities: Building Awareness and Ownership of Sequestration Projects

Local communities are at the heart of successful carbon sequestration efforts. By involving residents in the design and implementation of projects, governments and organizations can build lasting commitment to environmental stewardship. When community members understand the importance of sequestration projects, they become active participants rather than passive beneficiaries. This sense of ownership encourages the adoption of sustainable practices, such as regenerative agriculture and agroforestry, which enhance carbon capture while providing local economic and ecological benefits.

In countries like Kenya, local engagement has helped to scale up carbon farming initiatives. Through community-based education programs, farmers adopt practices that increase soil carbon levels, such as reduced tillage, mulching, and cover cropping. These approaches improve both the soil's capacity to store carbon and the resilience of local agriculture to climate impacts (FAO, 2024).

Case Study: Kenya's Success in Training 5,000 Farmers in Sustainable Practices

One of the most successful examples of community engagement in carbon sequestration is the **Kenya Agricultural Carbon Project (KACP)**. Launched in partnership with the World Bank and Vi Agroforestry, this project has trained **over 5,000 smallholder farmers** in sustainable land management practices. Farmers learned techniques such as agroforestry, composting, and crop rotation, which not only sequester carbon but also improve crop yields and boost soil health. The project has enabled farmers to sequester significant amounts of carbon, earning carbon credits through the Verified Carbon Standard (VCS) system (World Bank, 2014).

The success of KACP lies in its holistic approach, combining practical training with financial incentives. Farmers receive payments for carbon sequestration, providing them with an additional income stream that encourages the continued adoption of sustainable practices. The project's impact is significant, contributing to local livelihoods while enhancing the region's capacity to capture and store carbon (Verra, 2017).

Empowering Through Education: How Integrating Local Knowledge Drives Community Success

Education is a powerful tool for empowering communities to take climate action. Integrating local knowledge into carbon sequestration projects allows initiatives to adapt to the unique environmental and cultural context of a region. In Kenya, for example, modern training programs that teach sustainable farming techniques have enhanced traditional agricultural practices. This combination of Indigenous knowledge and new methodologies has led to a more resilient and sustainable agricultural system (FAO, 2024).

Programs that prioritize education not only improve the effectiveness of sequestration projects but also foster long-term commitment. As community members gain the skills and knowledge to manage these projects, they are better equipped to pass on sustainable practices to future generations, ensuring the longevity of carbon capture initiatives.

Conclusion

Community engagement and education are critical components of successful carbon sequestration projects. By empowering residents with knowledge and fostering a sense of ownership, governments and organizations can ensure that these initiatives are both impactful and sustainable. Kenya's success in training farmers to adopt carbon-sequestering practices highlights the transformative potential of education-driven climate action.

Key Takeaway

Engaging local communities and integrating education into carbon sequestration initiatives fosters long-term sustainability and amplifies the success of carbon capture projects, as evidenced by Kenya's model of training smallholder farmers in sustainable practices.

Chapter 9
Measuring and Verifying Carbon Sequestration

Carbon sequestration plays a vital role in reducing atmospheric CO2, but its effectiveness depends on accurate measurement and verification. Ensuring that carbon sequestration practices deliver the promised environmental benefits requires sophisticated tools and robust systems of validation. This chapter explores key technologies for measuring carbon capture, highlights a case study of California farmers successfully entering the carbon market, and discusses the importance of verification in maintaining transparency and trust within carbon markets.

Tools for Accurate Measurement: Remote Sensing, Soil Sampling, and Carbon Credit Quantification

Accurate measurement of carbon sequestration is essential for quantifying its impact and ensuring that carbon credits are valid. **Remote sensing** technology, often utilizing satellite data, has revolutionized the ability to monitor carbon storage in soil and vegetation. This technology allows large areas of land to be assessed efficiently and cost-effectively, offering a solution to the logistical challenges of traditional soil sampling (Boomitra, 2023). In addition to remote sensing, **soil sampling** remains a crucial method for verifying carbon levels at a more granular level. By collecting and analyzing soil samples from different sites, scientists can measure the actual carbon content and ensure that sequestration practices are working as intended (AFT, 2024).

Another important tool in carbon sequestration is the quantification and validation of **carbon credits**. Farmers and landowners who implement sustainable practices such as no-till farming or agroforestry can sell carbon credits in the voluntary carbon market, receiving financial compensation for the carbon they store. Verification of these credits often requires third-party validation, where auditors ensure that carbon is being sequestered in accordance with the terms of the agreement (FoodChain ID, 2024).

Case Study: California Farmers' Success in Entering Carbon Markets with Verified Credits

California farmers have embraced carbon sequestration through sustainable farming practices, leading to their successful entry into carbon markets. Companies like

Boomitra have developed AI-powered platforms to measure carbon sequestration remotely, making it easier for farmers to quantify their carbon storage and sell verified credits (Boomitra, 2023). In collaboration with platforms like **Athian**, California dairy farmers have also leveraged carbon insetting markets to reduce methane emissions and create verified carbon credits (Elanco, 2024). These verified credits are then sold to companies looking to offset their emissions, providing a new revenue stream for farmers while contributing to climate mitigation.

One key success factor for these farmers is the transparency and accuracy provided by remote sensing and soil sampling, which have made it possible to generate verified credits. With these credits, farmers are rewarded not just for sustainable practices but also for their contributions to global climate goals.

The Importance of Verification: Ensuring Accurate and Transparent Results in Carbon Markets

Verification is the cornerstone of trust in carbon markets. Without reliable validation of carbon credits, the integrity of these markets can be compromised. Third-party verification bodies like **FoodChain ID** play a critical role in ensuring that carbon sequestration practices are implemented correctly and that the resulting carbon credits reflect genuine carbon reductions (FoodChain ID, 2024). These organizations use a combination of remote sensing data, soil sampling, and compliance audits to validate the carbon stored in ecosystems and agricultural soils.

Moreover, verification helps maintain the transparency needed to attract investors and ensure that corporations are truly offsetting their emissions by purchasing legitimate carbon credits. This layer of accountability ensures that carbon sequestration initiatives are both effective and sustainable in the long term.

Conclusion

The future of carbon sequestration lies in our ability to measure and verify its impact accurately. Tools like remote sensing, soil sampling, and third-party verification systems are critical for ensuring that carbon credits reflect real-world reductions in atmospheric CO_2. California farmers' success in entering the carbon markets through verified credits highlights the importance of these technologies in creating financial incentives for sustainable practices.

Key Takeaway

Accurate measurement and verification of carbon sequestration are essential for maintaining the integrity of carbon markets. Tools like remote sensing and soil sampling, combined with third-party verification, enable farmers and landowners to earn verified carbon credits while contributing to global climate solutions.

Chapter 10
Challenges and Future Directions

As carbon sequestration initiatives continue to expand, they face significant challenges that need to be addressed to maximize their impact. From funding gaps and technological limitations to land-use conflicts, overcoming these obstacles is critical for scaling up efforts. In this chapter, we examine the barriers to effective carbon sequestration, explore emerging trends like blockchain technology in carbon trading, and provide a call to action for continued advocacy and participation in local sequestration efforts.

Overcoming Barriers: Addressing Funding Gaps, Technological Limitations, and Land-Use Conflicts

One of the most persistent challenges facing carbon sequestration projects is the **funding gap**. Large-scale sequestration initiatives, whether in agriculture, forestry, or industrial sectors, require substantial financial investments. Governments and private investors alike often hesitate to fund these projects because of uncertainties around their long-term viability and the complexities of measuring and verifying carbon storage. For example, many farmers lack the initial capital to adopt sustainable practices like no-till farming or agroforestry, even though these techniques increase soil carbon sequestration (McKinsey, 2023).

Technological limitations also pose a significant barrier. While advances like bio-energy with carbon capture and storage (BECCS) and direct air capture (DAC) are promising, these technologies are still in the early stages of deployment. Costs remain high, and scaling these innovations to meet global climate goals requires further development and investment. These technologies face significant challenges in storing the captured carbon safely and permanently.

Another key obstacle is **land-use conflict**. Many sequestration initiatives, particularly those involving reforestation or wetland restoration, require significant land areas. However, this can clash with other land-use priorities such as agriculture, housing development, or industrial expansion. Navigating these competing interests requires careful planning and the involvement of local communities to ensure that land-use decisions reflect the broader needs of society and the environment.

Emerging Trends: The Role of Blockchain in Carbon Trading and Transparency

One of the most exciting emerging trends in carbon sequestration is the use of **blockchain technology** to enhance transparency and efficiency in carbon trading. Blockchain—a decentralized and secure ledger system—allows for the accurate recording and tracking of carbon credits, ensuring that each transaction is transparent and traceable. This can help address one of the most significant challenges in carbon markets: the lack of trust in the validity of carbon credits (UNFCCC, 2021).

By using blockchain, carbon sequestration projects can record data in real time, allowing for the verification of carbon storage with unprecedented accuracy. This technology also has the potential to reduce transaction costs and streamline the carbon trading process, making it easier for smallholders, farmers, and businesses to take part in the carbon market. For example, IBM and Energy Blockchain Lab are already working on blockchain platforms to facilitate carbon asset trading in China, offering a glimpse of how this technology could revolutionize global carbon markets (UNFCCC, 2021).

Blockchain could also help solve issues related to **double-counting** in carbon markets, where multiple entities claim the same carbon credit. Blockchain creates a transparent and immutable record of each transaction, ensuring carbon credits are only counted once, which boosts the integrity of carbon markets and attracts more investors to fund sequestration projects (McKinsey, 2023).

Call to Action: Encouraging Advocacy and Participation in Local Sequestration Efforts

Addressing the challenges of carbon sequestration will require collective action at both local and global levels. **Advocacy** plays a crucial role in ensuring that governments continue to fund and support these projects, particularly in regions where sequestration efforts can provide significant environmental and economic benefits. By raising awareness and pushing for more robust climate policies, individuals and organizations can help close the funding gap and expand the reach of these initiatives.

Community participation is key to the success of localized carbon sequestration efforts. Farmers, landowners, and local governments need to be directly involved in the planning and implementation of these projects to ensure their sustainability. Local stakeholders bring valuable knowledge about their environments and are best positioned to identify solutions that work for their specific regions. Encouraging community engagement in sequestration projects not only improves their effectiveness but also fosters a sense of ownership and responsibility.

Conclusion

The future of carbon sequestration hinges on our ability to overcome key challenges, such as funding gaps, technological limitations, and land-use conflicts. However, emerging trends like blockchain technology offer exciting possibilities for improving transparency and efficiency in carbon markets, making it easier to scale these efforts. By advocating for continued support and encouraging local participation, we can ensure that carbon sequestration becomes a cornerstone of global climate action.

Key Takeaway

Overcoming the barriers to carbon sequestration requires innovative solutions, such as blockchain for transparency in carbon markets, alongside collective action and community participation to ensure the success of localized efforts.

Conclusion
The Path Forward

The journey to mitigate climate change through carbon sequestration is ambitious, but it is one we can collectively undertake by focusing on local actions that lead to global impact. Throughout this book, we've explored a variety of approaches, from agricultural practices and urban initiatives to innovative technologies and policy frameworks, all aimed at capturing and storing carbon. This conclusion revisits the key insights gained and provides practical steps for readers to support carbon sequestration efforts within their communities, highlighting the transformative power of localized actions.

Key Insights Recap: Highlighting Critical Takeaways from Each Chapter

The critical insights shared throughout this book reflect the diversity of methods available to sequester carbon. In **Chapter 1**, we laid the foundation by discussing the importance of carbon sequestration and the various methods, such as biological and geological sequestration, that contribute to reducing atmospheric CO_2. **Chapter 2** explored how localized approaches, tailored to specific environments, enhance the effectiveness of carbon sequestration, with case studies from the Midwest U.S. and rural India demonstrating significant gains in soil carbon levels.

Chapter 3 focused on agricultural practices, such as no-till farming and agroforestry, which not only sequester carbon but also improve soil health and crop yields. The success of

Brazilian farmers increasing soil organic carbon by 1.5% annually stands out as a powerful example of the dual benefits these techniques offer. In **Chapter 4**, we examined urban initiatives, including green roofs, urban forests, and community gardens, which provide additional benefits like improving air quality and urban biodiversity while capturing CO2.

In **Chapter 5**, we saw the potential of ecosystem restoration, with the rewilding of the Scottish Highlands sequestering millions of tons of CO2, a testament to the power of natural landscapes. **Chapter 6** introduced us to innovative technologies like BECCS and DAC, which are pushing the boundaries of carbon capture. **Chapters 7 through 9** reinforced the need for supportive policies, robust community engagement, and transparent measurement tools like blockchain to drive these efforts forward. Together, these insights underscore the breadth of opportunities available to tackle climate change through carbon sequestration.

Local Actions for Global Impact: Practical Steps for Readers to Support Carbon Sequestration in Their Communities

The success stories and strategies outlined in this book highlight one clear truth: local actions have the potential to create significant global change. Here are practical steps readers can take to support carbon sequestration in their communities:

1. **Engage in Sustainable Agriculture**: Whether you own land or support local farmers, advocating for and adopting practices like no-till farming, cover cropping, and agroforestry can enhance soil carbon storage while improving yields and resilience.

Volunteering with local agricultural groups or supporting carbon farming initiatives can also make a direct impact.

2. **Participate in Urban Greening**: Advocate for or initiate urban greening projects such as planting trees, creating community gardens, or installing green roofs. These projects not only sequester carbon but also improve air quality and provide green spaces for community well-being.

3. **Support Ecosystem Restoration**: Participate in local reforestation or wetland restoration projects, or donate to organizations that work on restoring natural ecosystems. These initiatives sequester carbon, restore biodiversity, and combat the effects of land degradation.

4. **Advocate for Carbon Sequestration Policies**: Local governments play a crucial role in supporting sequestration efforts. By advocating for policies that incentivize carbon farming, urban greening, and renewable energy, individuals can help shape the regulatory landscape to support these initiatives.

5. **Get Involved in Carbon Markets**: Farmers and landowners can participate in carbon markets by earning credits for sequestering carbon through verified practices. Engage with platforms that support transparency in carbon markets and promote community-led projects.

The Power of Localized Efforts: Reinforcing the Idea That Community-Level Initiatives Can Drive Significant Global Change

At the heart of this book is the belief that **localized efforts have the power to drive global change**. Each community, no matter how small, has the potential to contribute to the global fight against climate change by adopting practices that sequester carbon. Local actions, whether in agriculture, urban planning, or ecosystem restoration, not only reduce emissions but also create resilient communities better equipped to face climate challenges.

The stories shared in this book—whether of Brazilian farmers increasing soil carbon, Melbourne's urban greening projects, or Kenya's successful farmer training programs—demonstrate how local actions can accumulate into significant global impacts. By acting collectively, from the grassroots up, we can transform how we address climate change, ensuring a sustainable future for generations to come.

Conclusion

The path forward for carbon sequestration is clear: local actions, supported by innovative technologies, strong policies, and community engagement, have the potential to reshape our planet's carbon footprint. Each of us has a role to play, whether through supporting sustainable agriculture, advocating for urban greening, or participating in carbon markets. The power of localized efforts should not be underestimated—together, we can drive global change.

Key Takeaway

Localized actions in carbon sequestration—whether through sustainable farming, urban greening, or ecosystem restoration—have the potential to collectively drive global change. By engaging at the community level, we can all contribute to a more sustainable and resilient planet.

Frequently Asked Questions

This section provides straightforward answers to common questions about carbon sequestration, carbon credits, project implementation, and how individuals can contribute to reducing CO2 emissions. Whether you're a beginner or someone familiar with the topic, these FAQs help you engage with carbon sequestration efforts on a personal level.

1. What is carbon sequestration?

Carbon sequestration is capturing and storing atmospheric carbon dioxide (CO_2) in natural or artificial reservoirs. This helps reduce the amount of CO_2 in the atmosphere, mitigating the greenhouse effect and slowing climate change. It can occur naturally in forests, soils, and oceans, or through human-made technologies like carbon capture and storage (CCS).

2. How do carbon credits work?

Carbon credits are tradable certificates representing one metric ton of CO_2 (or its equivalent) removed from or avoided in the atmosphere. Individuals or companies that reduce or offset their carbon emissions through verified carbon sequestration projects can earn credits. Sellers can trade these credits on the voluntary or compliance carbon market, where organizations purchase them to offset their emissions and meet sustainability goals.

3. How can I take part in carbon markets?

If you manage land or own a business involved in sustainable practices like reforestation, regenerative farming, or renewable energy, you may be eligible to earn carbon credits. Taking part in carbon markets typically involves getting a project verified by a third-party organization, such as Verra or Gold Standard, which ensures your sequestration efforts are quantifiable and meet market standards. Once verified, you can sell the credits to buyers in the carbon market.

4. What are the costs of implementing carbon sequestration projects?

The cost of implementing a carbon sequestration project varies widely based on the method and scale. For example, reforestation may have lower upfront costs compared to more technological solutions like direct air capture (DAC). Agriculture-based projects, like no-till farming or agroforestry, involve initial investments in equipment and labor but often result in long-term financial and environmental benefits. Grants, government subsidies, and carbon credit earnings can help offset these costs.

5. How much CO2 do common practices sequester?

The method used determines how much CO2 is sequestered. Reforestation projects can sequester an average of 10-20 metric tons of CO2 per hectare annually, while no-till farming can sequester approximately 0.1-0.5 metric tons of CO2 per acre each year. Advanced technologies like DAC can capture up to 1 million tons of CO2 annually per facility, though they are more resource-intensive.

6. How can I, as an individual, contribute to reducing CO2 emissions?

Individuals can contribute to carbon sequestration in several ways:

- **Plant trees**: Support local reforestation efforts or start your own tree-planting projects.

- **Adopt sustainable gardening or farming practices**: Incorporate no-till gardening, composting, and cover cropping to improve soil carbon levels.

- **Support verified carbon offset programs**: Many organizations offer opportunities for individuals to purchase carbon credits that support projects around the world.

- **Advocate for local sequestration projects**: Engage with local governments or community groups to promote urban greening, community gardens, or policy changes that support carbon capture.

- **Reduce personal carbon footprint**: Minimize energy use, drive less, and reduce waste to lower your individual emissions and help offset atmospheric CO_2.

7. What is the difference between biological and geological sequestration?

Biological sequestration refers to the natural process of storing carbon in living organisms like plants, soil, and marine ecosystems. Examples include reforestation, wetland restoration, and agricultural practices that improve soil health. Geological sequestration involves capturing CO_2 emissions from industrial processes and storing them underground in rock formations, such as depleted oil fields or saline aquifers. Geological sequestration typically uses human-made technology, while biological sequestration relies on natural processes.

8. What role do governments play in supporting carbon sequestration?

Governments provide crucial support for carbon sequestration efforts through regulatory frameworks, financial incentives, and policy initiatives. Programs like carbon taxes, subsidies for sustainable farming, and grants for reforestation projects help reduce financial barriers for individuals and businesses. Additionally, governments may implement carbon pricing mechanisms, such as cap-and-trade systems, which encourage organizations to invest in carbon reduction and sequestration.

9. Can carbon sequestration projects help generate income?

Yes, carbon sequestration projects can generate income through the sale of carbon credits in voluntary and compliance markets. Landowners, farmers, and businesses that implement verifiable sequestration practices, such as reforestation or soil carbon storage, can sell credits to organizations looking to offset their emissions. This provides a financial incentive to continue investing in sustainable land management practices.

10. What is the future of carbon sequestration?

The future of carbon sequestration is promising, with emerging technologies like bioenergy with carbon capture and storage (BECCS) and direct air capture (DAC) expected to play a larger role in global climate strategies. As carbon markets expand and governments implement stricter emissions targets, sequestration projects will probably become more common and profitable. However, we must address challenges like funding, technological scalability, and land-use conflicts to ensure long-term success.

Case Study Summary

This section provides readers with a concise overview of the key case studies featured in *Localized Carbon Sequestration Success Stories*. Each case study highlights successful carbon sequestration efforts across different regions and sectors, offering a quick reference for practical applications of carbon capture and storage techniques.

---

1. Midwest U.S. – Soil Carbon Enhancement through No-Till Farming

Region: Midwest, United States

Sector: Agriculture

Sequestration Method: No-till farming, cover cropping

Summary: In the Midwest, farmers adopting no-till farming and cover cropping techniques have seen a 20%

increase in soil carbon levels over five years. These practices reduce soil disturbance, allowing organic matter to build up and store carbon in the soil. By minimizing tillage, farmers also conserve water and reduce erosion, further improving the land's ability to sequester carbon.

---

2. Brazil – Regenerative Agriculture Increasing Soil Organic Carbon

Region: Brazil

Sector: Agriculture

Sequestration Method: Regenerative agriculture

Summary: Brazilian farmers have successfully increased soil organic carbon by an average of 1.5% annually through regenerative agricultural practices, such as crop rotation, agroforestry, and organic soil amendments. These efforts have not only enhanced carbon sequestration but also improved soil health and biodiversity on degraded land, showing the dual benefits of sustainable farming.

---

3. Scottish Highlands – Rewilding for Ecosystem Restoration

Region: Scottish Highlands

Sector: Ecosystem Restoration

Sequestration Method: Reforestation and rewilding

Summary: The rewilding project in the Scottish Highlands has led to the sequestration of 5 million tons of CO_2 over the past 15 years. By planting native trees and restoring natural habitats, the project has successfully revived the Caledonian Forest and enhanced biodiversity. The reintroduction of species and natural regeneration has increased the landscape's ability to capture carbon while supporting local economies through eco-tourism.

4. Melbourne, Australia – Urban Greening for Carbon Sequestration

Region: Melbourne, Australia

Sector: Urban Initiatives

Sequestration Method: Green roofs, urban forests

Summary: Melbourne's urban greening projects, including the development of green roofs, urban forests, and community gardens, have sequestered approximately 300,000 tons of CO_2 over the last decade. These initiatives improve air quality, reduce the urban heat island effect, and foster biodiversity in the city while contributing to carbon reduction efforts.

5. Kenya – Sustainable Farming Training for Carbon Farming

Region: Kenya

Sector: Agriculture and Community Engagement

Sequestration Method: Agroforestry, reduced tillage, sustainable farming practices

Summary: The Kenya Agricultural Carbon Project (KACP) has trained over 5,000 smallholder farmers in sustainable land management practices. These farmers have adopted agroforestry, reduced tillage, and composting to enhance soil carbon storage. The project has also allowed farmers to earn carbon credits, contributing to their livelihoods while improving the resilience of local ecosystems.

6. California, United States – Carbon Markets for Verified Credits

Region: California, United States

Sector: Agriculture and Carbon Markets

Sequestration Method: Soil carbon capture and remote sensing

Summary: California farmers have successfully entered carbon markets through soil carbon capture verified by remote sensing technologies. By adopting sustainable practices like no-till farming and crop rotation, these farmers earn carbon credits that are sold on the voluntary carbon market. Their participation has contributed to increased carbon sequestration while providing an additional revenue stream for sustainable farming.

7. South Texas, United States – Direct Air Capture (DAC) Hub

Region: South Texas, United States

Sector: Technology

Sequestration Method: Direct air capture (DAC)

Summary: The South Texas DAC Hub is part of a $1.2 billion federal initiative and is expected to remove up to 1 million tons of CO_2 annually by 2025. The facility captures CO_2 directly from the atmosphere and stores it in underground reservoirs, making it one of the largest DAC projects in the U.S. This project highlights the potential of cutting-edge technology to address carbon emissions on a large scale.

Resources

The resources provided in this section are designed to equip you with the knowledge, tools, and connections necessary to further explore and participate in carbon sequestration efforts. Whether you're a researcher, policymaker, farmer, or community leader, these curated resources offer practical insights, scientific data, and opportunities to engage with carbon reduction initiatives at various levels.

From groundbreaking scientific studies and technological platforms to government-backed programs and community-based organizations, each resource has been selected to enhance your understanding of localized carbon sequestration. These materials not only expand on the topics covered in this book but also provide actionable steps and real-world applications to help you implement sustainable practices in your own community.

By diving into these resources, you'll be empowered to take part in the global movement toward carbon sequestration and climate resilience, making a tangible difference in the fight against climate change.

1. Scientific Journals and Publications

Nature Climate Change: A leading journal that publishes the latest research on climate change, including carbon sequestration techniques and their effectiveness.

Global Change Biology: Covers studies on the biological aspects of carbon sequestration in forests, wetlands, and agricultural systems.

Environmental Research Letters: A source for peer-reviewed studies on environmental science, with specific attention to carbon capture, storage, and climate change mitigation strategies.

2. Organizations and NGOs

International Union for Conservation of Nature (IUCN): Provides resources and case studies on ecosystem restoration and its role in carbon sequestration.

The Nature Conservancy: Offers insights into practical carbon sequestration efforts, such as reforestation and agricultural projects, and tools for community engagement.

Carbon Tracker Initiative: Analyzes the impact of carbon sequestration on markets and explores the potential for investment in carbon capture technologies.

3. Government Agencies

US Environmental Protection Agency (EPA): Comprehensive resources on carbon sequestration, government-backed initiatives, funding opportunities, and policy frameworks.

European Commission: Information on the European Green Deal, carbon farming policies, and sustainability investments.

United Nations Framework Convention on Climate Change (UNFCCC): Valuable information on international carbon markets, climate finance, and sustainable development practices.

4. Carbon Credit Platforms

Verra: A global leader in the voluntary carbon market, offering methodologies and standards for carbon sequestration projects and carbon credit verification.

Gold Standard: Sets rigorous benchmarks for carbon reduction projects and provides resources on how communities can generate verified carbon credits.

ClimeCo: Offers tools for landowners and organizations to take part in carbon trading markets and manage their carbon footprint.

5. Books and Educational Resources

Drawdown: The Most Comprehensive Plan Ever Proposed to Reverse Global Warming by Paul Hawken: Outlines various climate solutions, including carbon sequestration, grounded in science and practicality.

Soil and Carbon: Soil Health, Climate Change, and Carbon Sequestration by Ronald Amundson: Focuses on the role of soils in capturing carbon and agricultural practices.

Rewilding by Paul Jepson and Cain Blythe: Explores the contribution of rewilding to carbon sequestration through ecosystem restoration.

6. Technological Platforms and Tools

Carbon Mapper: Uses satellite technology to track and measure methane and CO2 emissions, providing data for carbon sequestration initiatives.

Terraton Initiative: Helps farmers take part in carbon sequestration projects through sustainable agricultural practices and provide tools to measure soil carbon levels.

Boomitra: An AI-driven platform that helps monitor soil carbon levels and create verified carbon credits for trading.

7. Community-Based Resources

Kiss the Ground: Promotes regenerative agriculture and soil health education through training programs and community workshops.

Farmers for Climate Action: Connects farmers with resources to implement climate-smart agriculture and carbon farming practices.

Urban Green Council: Provides resources and case studies on urban greening initiatives aimed at reducing carbon emissions in cities.

Glossary

Agroforestry: A land-use management system integrates trees and shrubs into agricultural fields to enhance carbon sequestration, improve soil health, and increase biodiversity.

BECCS (Bioenergy with Carbon Capture and Storage): A technology that combines biomass energy production with carbon capture and storage to remove CO_2 from the atmosphere while generating energy.

Biological Sequestration: capturing and storing carbon in living organisms, such as plants, soil, and marine ecosystems, through natural biological processes like photosynthesis and soil carbon storage.

Carbon Credits: Tradable certificates representing the removal or reduction of one metric ton of CO_2 or its equivalent from the atmosphere, used to offset emissions in carbon markets.

Carbon Farming: Agricultural practices specifically designed to capture and store carbon in soil and vegetation, which helps mitigate climate change while improving soil health and crop yields.

Carbon Offset: A reduction in CO_2 emissions made to compensate for emissions elsewhere, often achieved through the purchase of carbon credits from verified sequestration projects.

Carbon Sequestration: capturing and storing atmospheric CO_2 in natural or artificial reservoirs, such as soils, forests, or underground geological formations, to mitigate climate change.

Carbon Trading: A market-based system where companies or entities buy and sell carbon credits to meet regulatory limits on CO_2 emissions or to offset their carbon footprint voluntarily.

DAC (Direct Air Capture): A technology directly removes CO_2 from the atmosphere using chemical processes, allowing industries to store or use the captured carbon.

Geological Sequestration: storing captured CO_2 underground in geological formations, such as depleted oil and gas reservoirs or deep saline aquifers, to prevent its release into the atmosphere.

No-Till Farming: An agricultural practice where the soil is left undisturbed by tillage, helping to preserve soil carbon, reduce erosion, and improve water retention, thereby enhancing carbon sequestration.

Paris Agreement: An international treaty signed by nearly 200 countries in 2015 to limit global warming to below 2°C, with a goal of keeping temperature rise under 1.5°C compared to pre-industrial levels.

Rewilding: The process of restoring ecosystems to their natural state by reintroducing native species and encouraging natural processes like forest regeneration, which aids in carbon sequestration.

Soil Carbon: Organic carbon stored in the soil, derived from decomposed plant and animal matter. Soil carbon plays a vital role in carbon sequestration and maintaining healthy ecosystems.

Sustainable Investments: Financial investments that focus on generating social and environmental benefits alongside financial returns, often supporting projects that promote sustainability and carbon sequestration.

Urban Greening: The practice of incorporating more vegetation, such as trees, green roofs, and community gardens, into urban areas to improve air quality, reduce heat, and sequester carbon.

Verification (Carbon Markets): The process of independently assessing and confirming that a project's carbon sequestration practices are valid and that the corresponding carbon credits accurately reflect the amount of CO_2 captured or offset.

Verified Carbon Standard (VCS): A certification program that ensures carbon credits are valid and meets stringent environmental and social criteria for carbon reduction projects.

References

American Farmland Trust (AFT). (2024). A guide to agricultural carbon markets and sustainable farming practices. *AFT News*.

Bachu, S. (2008). CO2 storage in geological media: Role, means, status, and barriers to deployment. *Progress in Energy and Combustion Science*, 34(2), 254-273.

BCG. (2024). Shifting the Direct Air Capture Paradigm: Can DAC Technology Scale in Time? Retrieved from www.bcg.com.

Benson, S. M., & Cole, D. R. (2008). CO2 sequestration in deep sedimentary formations. *Elements*, 4(5), 325-331.

Boomitra. (2023). How AI and satellite technology help farmers sequester carbon and sell verified carbon credits. *Caltech Magazine*.

Carbon Credits. (2024). Are Direct Air Capture plants facing massive clean energy challenge in the U.S.? Retrieved from www.carboncredits.com.

Carbon Herald. (2023). U.S. awards $1.2 billion for developing Direct Air Capture hubs in Texas and Louisiana. *Carbon Herald*. Retrieved from www.carbonherald.com.

City of Melbourne. (2021). Urban forests bloom across the city. *Melbourne News*.

Drax Group. (2020). The role of BECCS in decarbonizing industry. Retrieved from www.drax.com.

Elanco. (2024). Verified carbon credits in the dairy industry: A new revenue stream for sustainable farming. *Elanco Animal Health*.

European Commission. (2021). The European Green Deal: A plan to make Europe climate-neutral by 2050. Retrieved from https://ec.europa.eu.

European Commission. (2023). Common Agricultural Policy: Greening agriculture for a sustainable future. Retrieved from https://ec.europa.eu.

FAO. (2024). Sustainable agricultural practices in Kenya. *KBC News*.

FoodChain ID. (2024). Carbon credit verification: Ensuring transparency and accuracy in carbon markets. *FoodChain ID*.

Greening the West. (2021). Greening Melbourne: Reducing carbon and enhancing biodiversity.

Highlands Rewilding. (2021). Restoring the Scottish Highlands: A model for ecosystem restoration.

International Energy Agency. (2020). CCUS in Clean Energy Transitions. IEA. Retrieved from www.iea.org.

Lal, R. (2004). Soil carbon sequestration impacts on global climate change and food security. *Science*, 304(5677), 1623-1627.

McKinsey & Company. (2023). Carbon credits: Scaling voluntary markets. *McKinsey & Company*.

Oldfield, E., Bradford, M., & Wood, S. (2019). Global meta-analysis of the relationship between soil organic matter and crop yields. *Soil*, 5(1), 15-32.

Paustian, K., Larson, E., Kent, J., Marx, E., & Swan, A. (2019). Soil C sequestration as a biological negative emission strategy. *Frontiers in Climate*, 1, 8.

Rewilding Europe. (2021). Rewilding the Scottish Highlands.

Ramos, F. T., de C. Dores, E. F. G., dos S. Weber, O. L., & Beber, D. C. (2018). Soil organic matter doubles the cation exchange capacity of tropical soil under no-till farming in

Brazil. *Journal of the Science of Food and Agriculture*, 98(9), 3595-3602.

S&P Global. (2024). Direct Air Capture: Challenges and opportunities for clean power integration. Retrieved from www.spglobal.com.

Taghi Miri, T. (2024). Maximising CO2 sequestration in the city: The role of green walls in sustainable urban development. *Pollutants*, 4(1), 91-116.

UNFCCC. (2021). How blockchain technology could boost climate action. *UNFCCC*.

About the Author

Cassia Parham is a dedicated advocate for sustainability and environmental stewardship, with a focus on carbon sequestration and localized climate solutions. With extensive experience in promoting sustainable practices, Cassia blends a deep understanding of environmental science with a practical, results-driven approach to climate action.

Drawing from her background in environmental studies and a personal commitment to minimizing their ecological footprint, Cassia has become a trusted voice in the environmental movement. Known for translating complex environmental challenges into clear, actionable strategies, Cassia empowers communities and individuals to make impactful changes that contribute to global carbon reduction efforts.

Through insightful writing and an engaging, modern voice, Cassia Parham inspires readers to take meaningful action on climate issues. Beyond writing, Cassia enjoys regenerative gardening, exploring nature, and sharing knowledge on sustainable living practices that benefit both people and the planet.

www.ingramcontent.com/pod-product-compliance
Lightning Source LLC
Chambersburg PA
CBHW050308230526
45471CB00005B/2081